The LIBRARY of LANDFORMS™

GLACIERS

Isaac Nadeau

The Rosen Publishing Group's
PowerKids Press™
New York

To Emma and Sarah

Published in 2006 by The Rosen Publishing Group, Inc.
29 East 21st Street, New York, NY 10010

First Edition

Editor: Rachel O'Connor
Book Design: Elana Davidian

Photo Credits: Cover © Rich Reid/National Geographic/Getty Images; p. 4 © Jim DuFresne/Lonely Planet Images; p. 7 © Scott Darsney/Lonely Planet Images; p. 8 © Ralph Lee Hopkins/Lonely Planet Images; p. 11 (top) © Wolfgang Kaehler/Corbis; p. 11 (bottom) © Graeme Cornwallis/Lonely Planet Images; p. 12 (top) © Jim Clare/naturepl.com; p. 12 (bottom) © Maria Stenzel/National Geographic Image Collection; p. 15 © Jonathan Chester/Lonely Planet Images; p. 15 (inset) © Cheyenne Rouse - Braumberger/Lonely Planet Images; p. 16 (top) © Grant Dixon/Lonely Planet Images; p. 16 (bottom left) © Geoge D. Lepp/Corbis; p. 16 (bottom right) © Michael T. Sedam/Corbis; p. 19 © David Welling/naturepl.com; p. 20 (left) © Hal Horwitz/Corbis; p. 20 (right) © David Tipling/Lonely Planet Images.

Library of Congress Cataloging-in-Publication Data

Nadeau, Isaac.
Glaciers / Isaac Nadeau.— 1st ed.
p. cm. — (The library of landforms)
Includes index.
ISBN 1-4042-3124-2 (lib. bdg.)
1. Glaciers—Juvenile literature. I. Title. II. Series.
GB2403.8.N339 2006
551.31'2—dc22

2004025422

Manufactured in the United States of America

CONTENTS

As temperatures change throughout the seasons, glaciers also change. They grow larger with the winter snowfall. They get smaller in summer, when the heat melts some of the ice into water. When the ice melts a process called calving can occur. Calving happens when a large piece of ice falls away from the glacier causing a huge crash, as shown above in the Hubbard Glacier in Alaska.

What Is a Glacier?

When people hear the word "glacier," ice is the one thing that usually comes to mind. Ice is an important part of a glacier, but there is more to glaciers than just ice. People who study glaciers are called glaciologists. They say a glacier is a body of ice formed by snow that has been packed together. It gets packed over time under its own weight. To be called a glacier, this body of ice must also be in motion. It flows slowly across the land. Some glaciers are so large that entire mountains are buried in ice. Other glaciers are small, tucked away in a shaded hollow on a mountainside.

Many people do not realize that much of the land around us has been shaped by glaciers that moved across the land. **Landforms**, such as hills, **valleys**, and lakes, have been left behind from glaciers moving over land more than 10,000 years ago. In Alaska, where glaciers are still common, you can see the effects that glaciers are having on the land.

Ice Ages

For most of its 4.5-billion-year history, Earth has experienced changes in **temperature**. There were times when temperatures were warmer than they are today, and times when they were colder. The colder periods are called ice ages. During an ice age, temperatures are cold enough to cause glaciers to cover a big part of Earth's surface. The last ice age began between 2 and 3 million years ago. At its height temperatures on Earth were around 10°F (5.6°C) colder than average temperatures today.

An ice age can last for millions of years. There are colder and warmer periods within an ice age. Glaciers can grow and **shrink** with the changing temperatures. As recently as 10,000 years ago, all of Canada and Britain and much of the northern United States was buried beneath a huge mass of moving ice. At about this time, temperatures on Earth began to rise. These huge glaciers melted, **retreating** to the north where temperatures are cooler. Many glaciologists believe that we are still living in an ice age. They say in a few thousand years Earth's temperatures will become cooler and the glaciers will grow again.

Today many glaciers still exist on Earth. Some are tucked away high in the cold mountains of Alaska and California. Others are found near Earth's poles, like the Seuss Glacier shown here in Antarctica near the South Pole.

2.2 BILLION YEARS AGO: The earliest evidence of glaciers is found in rocks formed of till, a type of sediment deposited by glaciers.

600 MILLION YEARS AGO: The largest period of glaciation known to scientists begins. It is known as the Varangian glaciations.

2–3 MILLION YEARS AGO: The most recent ice age begins.

25,000 YEARS AGO: Ice sheets reach as far south as southern Illinois and central Missouri.

10,000 YEARS AGO: Ice sheets begin a major retreat as temperatures rise.

700–100 YEARS AGO: The Little Ice Age occurs, in which average temperatures drop and many glaciers advance worldwide.

TODAY: Glaciers cover about 10 percent of Earth's land surface, down from about 30 percent during the most recent glacial advance.

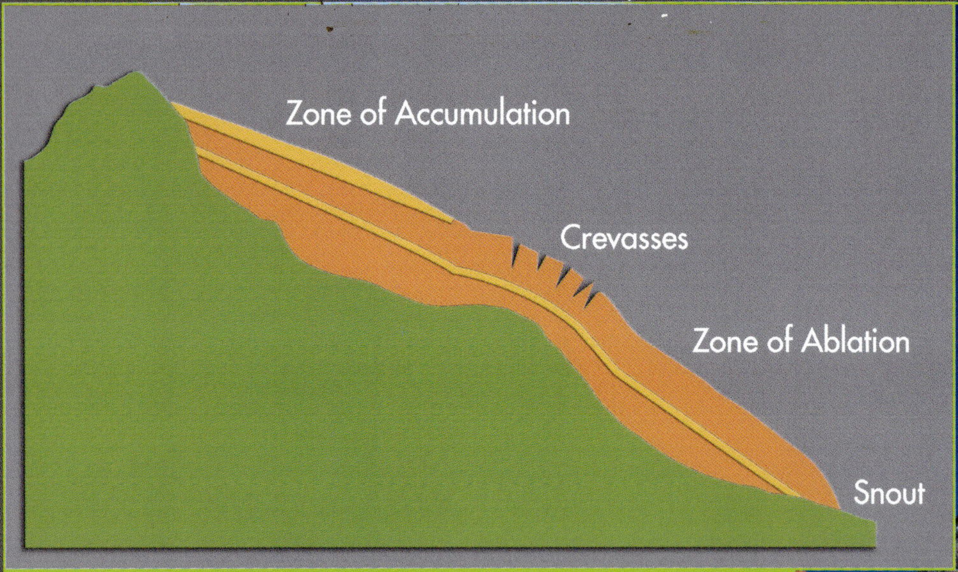

A glacier can be divided into two zones. In the zone of accumulation, more snow falls than melts each year. The zone of accumulation is located in the uphill, colder part of a glacier. The second area, called the zone of ablation, is where more snow melts than falls each year. The lowest end of a glacier, called the snout, is at the bottom of the zone of ablation. The snout is where water flows out of the glacier.

Shown here is a piece of glacier ice.

How Glaciers Are Formed

Glaciers form in cold **climates**. The coldest places on Earth are at the North Pole and the South Pole, which receive little direct sunlight, and near the tops of mountains.

It can take hundreds of years for a glacier to form. Glaciers form when the amount of snow that falls each winter is greater than the amount of snow that melts in the summer. The snow then piles higher and higher each year. As the pile grows, much of the air between the snowflakes is squeezed out. The snow becomes packed into hard little balls, called firn. Firn is about the size, shape, and hardness of grains of sand. When the firn is piled to depths of about 492 feet (150 m), the bottom layer is squeezed together into larger ice **crystals** called glacier ice. As more snow falls, more firn and glacier ice are formed. When the glacier ice reaches a thickness of about 150 feet (46 m), it begins to move under its own weight. It is pulled downhill by the force of **gravity**. When a mass of ice begins to move in this way, it is known as a glacier.

Glaciers can exist in areas with warm summers if enough snow falls in winter. In some areas, such as the Sierra Nevada in California, there is a huge amount of snowfall in the winter, but summers are very warm. Some places in the Sierra Nevada receive almost 40 feet (12 m) of snow each winter. This helps replace the snow that melted in the summer.

Ice Sheets and Ice Caps

Glaciers form in different sizes. The biggest glaciers of all are called ice sheets. There are only two glaciers on Earth that are large enough to be called ice sheets. Near the North Pole, the Greenland Ice Sheet covers about 80 percent of the country of Greenland. This ice sheet measures about 670,000 square miles (1.7 million sq km) and is up to 1.8 miles (2.9 km) thick. As big as the Greenland Ice Sheet is, it is small compared to the ice sheet that covers the **continent** of Antarctica at the South Pole. The Antarctic Ice Sheet is about 5.4 million square miles (14 million sq km) in area and is more than 2.5 miles (4 km) thick at its thickest.

Another type of glacier is an ice cap. Ice caps are smaller than ice sheets, but they can still cover thousands of square miles (sq km) and be more than 1,000 feet (305 m) thick. Ice caps are found in areas of Norway, Iceland, Asia, and Canada, as well as in Alaska. In these places the tops of tall mountains can often be seen poking out from the ice. These mountain peaks are called nunataks.

Top: By studying the ice in Antarctica, glaciologists believe that the Antarctic Ice Sheet was formed as early as 14 million years ago. Together the Greenland and Antarctic ice sheets make up almost 10 percent of the total land area on Earth. Shown here is an ice sheet in Antarctica. *Bottom:* You can see nunataks here as they peek out above the glaciers in the North-East Greenland National Park.

Top: Mount Kilimanjaro, the highest mountain in Africa at 19,333 feet (5,893 m), is only 220 miles (354 km) south of the equator. At its base far below, temperatures are high year-round, yet Kilimanjaro has a small glacier at its peak because temperatures there are often well below freezing. *Bottom:* Here you can see Taylor Glacier, a valley glacier in Antarctica. Like all glaciers, valley glaciers flow slowly downhill.

Mountain Glaciers

Not all glaciers are found close to the North Pole and the South Pole. Some are found in **temperate** regions, where temperatures average several degrees warmer than in polar regions. Glaciers in temperate regions have to form in high mountain ranges where the climate is cooler. These are called mountain glaciers. In North America many mountain glaciers can be found in Alaska and Canada. A few can be found in the Sierra Nevada of California and in the Colorado Rockies.

There are two types of mountain glaciers. They are valley glaciers and cirque glaciers. Valley glaciers are found in valleys. Valley glaciers appear as great rivers of ice filling up the valley floor. Valley glaciers are usually long and narrow. A cirque glacier is a small glacier found in a bowl-shaped hollow in the side of a mountain. Cirque glaciers are usually found at the tops of valleys, where valley glaciers used to be. In some cases a cirque glacier is all that is left of a much larger valley glacier that has melted away. Most cirque glaciers are so small that they do not have names.

How Glaciers Move

Glaciers melt when snowmelt in summer is greater than snowfall in winter. The huge ice sheets that covered northern North America 10,000 years ago retreated in this way. To the south temperatures were warmer, and so the southern parts of the glaciers melted more quickly than the northern parts.

There are two main ways in which a glacier can move. They are called creep and basal sliding. When glacial ice becomes about 150 feet (46 m) thick, the ice at the bottom becomes more **flexible**. When it is more flexible, it can flow slowly downhill under the force of gravity. This type of movement is called creep. For the most part, creep may occur at a rate of only a few feet (m) per year.

Basal sliding occurs when a glacier's bottom **layer** melts to become water. This layer of water reduces the **friction** between the ice and the rock below, and the glacier is able to slide more easily. There is more water at the bottom of a glacier during the summer because warmer temperatures cause the ice to melt. Therefore basal sliding usually happens more quickly during the summer than in the winter. In some cases a glacier can flow very quickly, as much as 150 feet (46 m) in a day. A glacier moving at this rate is said to be surging.

Glaciers, such as the one in Antarctica shown here, melt when climates become warmer. This is known as glacial retreat because the melting of the glacier makes it appear to be moving backward. In fact the glacier is just getting smaller. *Inset:* A crevasse is a crack that appears in a glacier as it creeps downhill. Here a climber explores a crevasse on Emmons Glacier in Washington, North America.

Top: A moraine is another landform created by glaciers. A moraine is a hill that is formed when till washes out of a glacier. Pictured here is the moraine-covered Khumbu Glacier in Nepal, India. *Bottom Left:* Here lakes are surrounded by eskers in the Northwest Territories of Canada. *Bottom Right:* This U-shaped valley in Washington State's North Cascades National Park was formed by a glacier.

How Glaciers Shape the Land

Glaciers shape the land through two basic methods. These are **erosion** and **deposition**. Erosion occurs as a glacier moves across the surface of the land. As a glacier moves, it scrapes the land beneath it. It also picks up bits of rock that range in size from tiny pieces of clay to huge boulders. Once these rocks are inside the ice, they help scratch whatever they come in contact with. In this way the land over which a glacier flows can become nearly flat. U-shaped valleys are formed when a glacier flows between mountains, eroding the sides of the mountains.

Deposition is the result of rocks being washed out with the water that **drains** from a melting glacier and being left behind on the ground or in the water. These rocks, called till, create many different kinds of landforms, including eskers and drumlins. An esker is a long, winding hill formed by till deposited in a river that flowed under the ice. A drumlin is formed when a glacier flows over a pile of till, shaping it into a dome.

Glaciers of California

There are hundreds of glaciers in California. Most of these are in the Sierra Nevada mountain range in the central part of the state. The glaciers of the Sierra Nevada were formed only about 700 years ago, during what is called the Little Ice Age. The Little Ice Age was a period of a few hundred years in which average temperatures on Earth were several degrees colder than they are today. Since California is in a temperate area, its glaciers experience warmer temperatures in summer than most of Earth's glaciers. Most of California's glaciers can be found in well-shaded areas, out of reach of much of the Sun's warmth. At about one-half square mile (1.3 sq km) in area, the Palisade Glacier is the largest glacier in the Sierra Nevada. It is located between 12,000 feet (3,658 m) and 13,400 feet (4,084 m) in **elevation**. An interesting ice formation called a moulin can be found on the Palisade Glacier. A moulin is a deep hole in the ice formed when water from the glacier's surface makes a depression in the ice. As more water flows into the depression it gets deeper and deeper. Finally a hole forms, like a well in the ice.

The Palisade Glacier in the East Sierra Mountains flows at a rate of about 23 feet (7 m) per year. Even as it flows forward, it is slowly shrinking, getting a little bit smaller year after year.

Left: Pictured here are yellow arctic poppies. *Right:* Here you can see a male emperor penguin keeping its chick warm by covering it with its belly. This is how the penguins keep their eggs warm, too.

Life On and Near Glaciers

Glaciers are cold places to live. Yet there are many kinds of plants and animals that make their homes in the regions where many glaciers can be found. In Greenland there are several kinds of mammals, including wolves, polar bears, and foxes. Some animals have thick coats of fur that keep them from freezing in winter. Others, such as the Arctic tern, migrate, or move to warmer climates during the cold winter. In Antarctica very few creatures can spend their entire lives on land. Many use the cold mountains as a place to raise their young. For example, emperor penguins come to Antarctica to lay their eggs during the winter months. The male keeps the eggs warm by keeping them on his feet, off the ice! However, many plants also live close to glaciers. Small plants such as mosses grow near glaciers. Flowers like the yellow arctic poppy and the purple sky pilot do, too. The glaciated mountains of Europe and North America are home to a tiny insect called the glacier flea. It lives on glaciers, feeding on **pollen** blown across the glacier by the wind.

Glaciers and People

In recent years scientists have discovered that Earth's temperatures are getting warmer. Glaciers around the world, from Kilimanjaro in Africa to the ice sheet in Greenland, are melting more rapidly than they have in a long time. Many scientists believe that this is in part caused by human activities such as the burning of oil and other fuels.

People such as the **Inuit** have been living near glaciers, in places such as Greenland and Canada, for thousands of years. Many people who live in northern climates make their living by hunting and fishing on the glacial ice.

In some places the water that melts from glaciers in summer is the main source of drinking water for a city or a village. Farmers also use water from glaciers to irrigate, or water, crops. Other places build dams to produce electricity using the energy created by glacial meltwater.

In Antarctica scientists drill deep holes in the ice and look at ice that was formed thousands of years ago. This ice provides clues about the amount of snowfall and the temperature of Antarctica in the distant past. It may help scientists forecast future ice ages, as well. As much as we know about glaciers, we still have a lot to learn about these important and beautiful landforms.

Glossary

climates (KLY-mits) The kinds of weather certain areas have.

continent (KON-tih-nent) One of Earth's seven large landmasses.

crystals (KRIS-tuhlz) Clear kinds of solids with angled and flat surfaces.

deposition (deh-puh-ZIH-shun) The dropping of tiny bits of rock in a new place.

drains (DRAYNZ) Causes a lessening in amount, content, power, or value.

elevation (eh-luh-VAY-shun) The height of an object.

erosion (ih-ROH-zhun) The wearing away of land over time.

flexible (FLEK-sih-bul) Being able to move and bend in many ways.

friction (FRIK-shin) The rubbing of one thing against another.

gravity (GRA-vih-tee) The natural force that causes objects to move toward Earth's center.

Inuit (IH-nyoo-it) Eskimos of North America.

landforms (LAND-formz) Features on Earth's surface.

layer (LEHR) One thickness of something.

pollen (PAH-lin) A powder made by the male parts of flowers.

retreating (rih-TREET-ing) The act of moving back.

shrink (SHRINGK) To make or become smaller because of heat, cold, or wetness.

temperate (TEM-puh-rit) Not too hot or too cold.

temperature (TEM-pruh-cher) How hot or cold something is.

valleys (VA-leez) Areas of low land between hills or mountains.

Index

Web Sites

Due to the changing nature of Internet links, PowerKids Press has developed an online list of Web sites related to the subject of this book. This site is updated regularly. Please use this link to access the list:
http://www.powerkidslinks.com/liblan/glaciers/